A GRIEF OBSERVED

BOOKS BY C. S. LEWIS

AVAILABLE FROM HARPERSANFRANCISCO

A Grief Observed

George MacDonald: An Anthology

Mere Christianity

Miracles

The Abolition of Man

The Great Divorce

The Problem of Pain

The Screwtape Letters (with "*Screwtape Proposes a Toast*")

The Weight of Glory

ALSO AVAILABLE FROM HARPERCOLLINS

The Chronicles of Narnia:

The Magician's Nephew

The Lion, the Witch and the Wardrobe

The Horse and his Boy

Prince Caspian

The Voyage of the Dawn Treader

The Silver Chair

The Last Battle

C. S. Lewis

A GRIEF OBSERVED

HarperSanFrancisco
A Division of HarperCollins*Publishers*

HarperCollins books may be purchased for educational, business, or sales promotional use. For information please write: Special Markets Department, HarperCollins Publishers, Inc., 10 East 53rd Street, New York, NY 10022.

HarperCollins Web site: http://www.harpercollins.com
HarperCollins®, ®, and HarperSanFrancisco™ are trademarks of HarperCollins Publishers, Inc.

HARPERCOLLINS EDITION 2001
Library of Congress Cataloging-in-Publication Data
Lewis, C. S. (Clive Staples), 1898–1963.
A grief observed / by C. S. Lewis. —HarperCollins ed.
p. cm.
ISBN 0-06-065238-1 (paperback)
1. Consolation. 2. Bereavement—Religious aspects—Christianity.
3. Davidman, Joy. 4. Lewis, C. S. (Clive Staples), 1898–1963—Religion.
I. Title.
BV4905.2.L4 2000
242'.4—dc21 00-063227

03 04 ❖/RRD 20 19 18 17 16 15 14

Contents

FOREWORD

When *A Grief Observed* was first published under the pseudonym of N. W. Clerk it was given me by a friend, and I read it with great interest and considerable distance. I was in the middle of my own marriage, with three young children, and although I felt great sympathy for C. S. Lewis in his grief over the death of his wife, at that time it was so far from my own experience that I was not deeply moved.

Many years later, after the death of my husband, another friend sent me *A Grief Observed* and I read it, expecting to be far more immediately involved than I had on the first reading. Parts of the book touched me deeply, but on the whole my experience of grief and Lewis's were very different. For one thing, when C. S. Lewis married Joy Davidman, she was in the hospital. He knew that he was marrying a woman who was dying of cancer. And even though there was the unexpected remission, and some good years of reprieve, his experience of marriage was only a taste, compared to my own marriage of forty years. He had been invited to the great feast of marriage and the banquet was rudely snatched away from him before he had done more than sample the hors d'oeuvres.

And to Lewis that sudden deprivation brought about a brief loss of faith. "Where is God? . . . Go to him when your need is desperate, when all other help is in vain, and what do you find? A door slammed in your face."

The death of a spouse after a long and fulfilling marriage is quite a different thing. Perhaps I have never felt more closely the strength of God's presence than I did during the months of my husband's dying and after his death. It did not wipe away the grief. The death of a beloved is an amputation. But when two people marry, each one has to accept that one of them will die before the other. When C. S. Lewis married Joy Davidman, it was a pretty certain expectation that she would die first, unless there was an unexpected accident. He moved into marriage with an imminent expectation of death, in an extraordinary witness of love and courage and personal sacrifice. Whereas a death which occurs after a full marriage and a reasonable life span is part of the whole amazing business of being born and loving and living and dying.

Reading *A Grief Observed* during my own grief made me understand that each experience of grief is unique. There are always certain basic similarities: Lewis mentions the strange feeling of fear, the needing to swallow, the forgetfulness. Perhaps all believing people feel, like Lewis, a horror of those who say of any tragedy, "Thy will be done," as though a God of love never wills anything but good for us creatures. He shows impatience with those who try to pretend that death is unimportant for the believer, an impatience which most of us feel, no matter how strong our faith. And C. S. Lewis and I share, too, the fear of the loss of memory. No photograph can truly recall the beloved's smile. Occasionally, a glimpse of someone walking down the street, someone alive, moving, in action, will hit with a pang of genuine recollection. But our memories, precious though they are, still are like sieves, and the memories inevitably leak through.

Like Lewis, I, too, kept a journal, continuing a habit started when I was eight. It is all right to wallow in one's journal; it is a way of getting rid of self-pity and self-indulgence and self-centeredness. What we work out in our journals we don't take out on family and friends. I am grateful to Lewis for the honesty of his journal of grief, because it makes quite clear that the human being is allowed to grieve, that it is normal, it is right to grieve, and the Christian is not denied this natural response to loss. And Lewis asks questions that we all ask: where do those we love go when they die?

Lewis writes that "I have always been able to pray for the dead, and I still do, with some confidence. But when I try to pray for H. [as he calls Joy Davidman in this journal], I halt." And this feeling I well understand. The beloved is so much a part of ourselves that we do not have the perspective of distance. How do we pray for what is part of own heart?

We don't have any pat answers. The church is still pre-Copernican in its attitude toward death. The medieval picture of heaven and hell hasn't been replaced with anything more realistic, or more loving. Perhaps for those who are convinced that only Christians of their own way of thinking are saved and will go to heaven, the old ideas are still adequate. But for most of us, who see a God of a much wider and greater love than that of the tribal God who only cares for his own little group, more is needed. And that more is a leap of faith, an assurance that that which has been created with love is not going to be abandoned. Love does not create and then annihilate. But where Joy Davidman is now, or where my husband is, no priest, no minister, no theologian can put into the limited terms of provable fact. "Don't talk to me about the consolations of religion," Lewis writes, "or I shall suspect that you do not understand."

For the true consolations of religion are not rosy and cozy, but com-forting in the true meaning of that word: com-fort: with strength. Strength to go on living, and to trust that whatever Joy needs, or anyone we love who has died needs, is being taken care of by that Love which began it all. Lewis rightly rejects those who piously tell him that Joy is happy now, that she is at peace. We do not know what happens after death, but I suspect that all of us still have a great deal to learn, and that learning is not necessarily easy. Jung said that there is no coming to life without pain, and that may well be true of what happens to us after death. The important thing is that we do not know. It is not in the realm of proof. It is in the realm of love.

I am grateful, too, to Lewis for having the courage to yell, to doubt, to kick at God with angry violence. This is a part of healthy grief not often encouraged. It is helpful indeed that C. S. Lewis, who has been such a successful apologist for Christianity, should have the courage to admit doubt about what he has so superbly proclaimed. It gives us permission to admit our own doubts, our own angers and anguishes, and to know that they are part of the soul's growth.

So Lewis shares his own growth and his own insights. "Bereavement is not the truncation of married love but one of its regular phases—like the honeymoon. What we want is to live our marriage well and faithfully through that phase, too." Yes, that is the calling of either husband or wife after the other has died.

I have pictures of my husband in my study, in my bedroom, now, after his death, as I had them around while he was alive, but they are icons, not idols; tiny flashes of reminders, not things in themselves, and, as Lewis says, sometimes a block rather than a help to the memory. "All reality is iconoclastic," he writes. "The earthly beloved, even in this life, incessantly triumphs over your mere idea of her. And you want her to; you want her with all her resistances, all her faults, all her unexpectedness. . . . And this, not an image or memory, is what we are to love still, after she is dead."

And that is more important than visitations from the dead, though Lewis discusses that possibility of these. In the end, what shines through the last pages of his journal of grief is an affirmation of love, his love for Joy and hers for him, and that love is in the context of God's love.

No easy or sentimental comforts are offered, but the ultimate purpose of God's love for all of us human creatures is love. Reading *A Grief Observed* is to share not only in C. S. Lewis's grief but in his understanding of love, and that is richness indeed.

Madeleine L'Engle
Crosswicks, August 1988

INTRODUCTION

A Grief Observed is not an ordinary book. In a sense it is not a book at all; it is, rather, the passionate result of a brave man turning to face his agony and examine it in order that he might further understand what is required of us in living this life in which we have to expect the pain and sorrow of the loss of those whom we love. It is true to say that very few men could have written this book, and even truer to say that even fewer men would have written this book even if they could, fewer still would have published it even if they had written it.

My stepfather, C. S. Lewis, had written before on the topic of pain (*The Problem of Pain*, 1940), and pain was not an experience with which he was unfamiliar. He had met grief as a child: he lost his mother when he was nine years old. He had grieved for friends lost to him over the years, some lost in battle during the First World War, others to sickness.

He had written also about the great poets and their songs of love, but somehow neither his learning nor his experiences had ever prepared him for the combination of both the great love and the great loss which is its counterpoint; the soaring joy which is the finding and winning of the mate whom God has prepared for us; and the crushing blow, the loss, which is Satan's corruption of that great gift of loving and being loved.

In referring to this book in conversation, one often tends to leave out, either inadvertently or from laziness, the indefinite article at the beginning of the title. This we must not do, for the title completely and thoroughly describes what this book is, and thus expresses very accurately its real value. Anything entitled "Grief Observed" would have to be so general and nonspecific as to be academic in its approach and thus of little use to anyone approaching or experiencing bereavement.

This book, on the other hand, is a stark recounting of one man's studied attempts to come to grips with and in the end defeat the emotional paralysis of the most shattering grief of his life.

What makes *A Grief Observed* even more remarkable is that the author was an exceptional man, and the woman whom he mourns, an exceptional woman. Both of them were writers, both of them were academically talented, both were committed Christians, but here the similarities end. It fascinates me how God sometimes brings people together who are so far apart, in so many ways, and merges them into that spiritual homogeneity which is marriage.

Jack (C. S. Lewis) was a man whose extraordinary scholarship and intellectual ability isolated him from much of mankind. There were few people among his peers who could match him in debate or discussion, and those who could almost inevitably found themselves drawn to one another in a small, tight-knit group which became known as "The Inklings," and which has left us with a legacy of literature. J.R.R. Tolkien, John Wain, Roger Lancelyn-Green, and Neville Coghill were among those who frequented these informal gatherings.

Helen Joy Gresham (née Davidman), the "H." referred to in this book, was perhaps the only woman whom Jack ever met who was his intellectual equal and also as well-read and widely educated as he was himself. They shared another common factor: they were both possessed of total recall. Jack never forgot anything he had read, and neither did she.

Jack's upbringing was a mixture of middle-class Irish (he came from Belfast, where his father was a police-court solicitor) and English, set in the very beginnings of the twentieth century—a time when the concepts of personal honour, total commitment to one's given word, and the general principles of chivalry and good manners were still drummed into the young British male with rather more intensity than was any other form of religious observance. The writing of E. Nesbit, Sir Walter Scott, and perhaps Rudyard Kipling were the exemplars of the standards with which Jack was indoctrinated as a young man.

My mother, on the other hand, could not have come from a background more divergent from his. The daughter of two lower-middle-class Jewish second-generation immigrants, her father of Ukrainian, her mother of Polish origins, she was born and brought up in the Bronx in New York City. The only striking similarities to be found in the comparison of their early developments were that they were both possessed of truly amazing intelligence combined with academic talent and eidetic memory. They both came to Christ via the long and difficult road which leads from Atheism, to Agnosticism, and thence by way of Theism finally to Christianity, and they both enjoyed remarkable success in their university student careers. Jack's was interrupted by his duty to his country in the First World War, and Mother's by political activism and marriage.

Much has been written, both fictional and factual (sometimes one masquerading as the other) concerning their lives and their meeting and marriage, but the most important part of the story pertaining to this book is simply a recognition of the great love that grew between them until it was an almost visible incandescence. They seemed to walk together within a glow of their own making.

To understand even a little of the agony which this book contains, and the courage it demonstrates, we must first acknowledge that love between them. As a child, I watched these two remarkable people come together, first as friends, then, in an unusual progression, as husband and wife, and finally as lovers. I was part of the friendship; I was an adjunct to the marriage, but I stood aside from the love. By that I do not mean that I was in any way deliberately excluded, but rather that their love was something of which I could not, and should not, be a part.

Even then in my early teen years I stood aside and watched the love grow between these two, and was able to be happy for them. It was a happiness tinged with both sadness and fear, for I knew, as did both Mother and Jack, that this, the best of times, was to be brief and was to end in sorrow.

I had yet to learn that all human relationships end in pain—it is the price that our imperfection has allowed Satan to exact from us for the privilege of love. I had the resilience of youth upon which to fall when Mother died; for me there would be other loves to find and no doubt in time to lose or be lost by. But for Jack this was the end of so much which life had for so long denied him and then briefly held out to him like a barren promise. For Jack there were none of the hopes (however dimly I might see them) of bright sunlit meadows and life-light and laughter. I had Jack to lean upon, poor Jack only had me.

I have always wanted the opportunity to explain one small thing that is in this book and which displays a misunderstanding. Jack refers to the fact that if he mentioned Mother, I would always seem to be embarrassed as if he had said something obscene. He did not understand, which was very unusual for him. I was fourteen when Mother died and the product of almost seven years of British Preparatory School indoctrination. The lesson I was most strongly taught throughout that time was that the most shameful thing that could happen to me would be to be reduced to tears in public. British boys don't cry. But I knew that if Jack talked to me about Mother, I would weep uncontrollably and, worse still, so would he. This was the source of my embarrassment. It took me almost thirty years to learn how to cry without feeling ashamed.

This book is a man emotionally naked in his own Gethsemane. It tells of the agony and the emptiness of a grief such as few of us have to bear, for the greater the love the greater the grief, and the stronger the faith the more savagely will Satan storm its fortress.

When Jack was racked with the emotional pain of his bereavement, he also suffered the mental anguish resulting from three years of living in constant fear, the physical agony of osteoporosis and other ailments, and the sheer exhaustion of spending those last few weeks in constant caring for his dying wife. His mind stretched to some unimaginable tension far beyond anything a lesser man could bear; he turned to writing down his thoughts and his reactions to them, in order to try to make some sense of the whirling chaos that was assaulting his mind. At the time that he was writing them, he did not intend that these effusions were to be published, but on reading through them some time later, he felt that they might well be of some help to others who were similarly afflicted with the turmoil of thought and feeling which grief forces upon us. This book was first published under the pseudonym of N. W. Clerk. In its stark honesty and unadorned simplicity the book has a power which is rare: it is the power of unabashed truth.

To fully appreciate the depths of his grief I think it is important to understand a little more of the circumstances of Jack and Mother's initial meeting and relationship. My mother and father (novelist W. L. Gresham) were both highly intelligent and talented people and in their marriage there were many conflicts and difficulties. Mother was brought up an atheist, and became a communist. Her native intelligence did not allow her to be deceived for long by that hollow philosophy, and (by this time, married to my father) she found herself searching for something less posturing and more real.

Encountering amid her reading of a wide variety of authors the work of the British writer C. S. Lewis, she became aware that beneath the fragile and very human veneer of the organized churches of the world, there lies a truth so real and so pristine that all of man's concocted philosophical posings tumble into ruin beside it. She became aware also that here was a mind of hitherto unparalleled clarity. As all new believers do, she had questions, and so she wrote to him. Jack noticed her letters at once, for they too signalled a remarkable mind, and a pen-friendship soon developed.

In 1952 Mother was working on a book about the Ten Commandments (*Smoke on the Mountain:* Westminster Press, 1953), and while convalescing from a serious illness journeyed to England determined to discuss the book with C. S. Lewis. His friendship and advice were unstinting as were those of his bother, W. H. Lewis, an historian and himself a writer of no mean ability.

On her return to America, Mother (now a complete Anglophile), discovered that her marriage to my father was over, and following the divorce she fled to England with myself and my brother. We lived for a while in London, and although letters were exchanged, Jack was not a visitor to our home, he rarely came to London, which was a city he was not fond of, and Mother and he were merely intellectual friends at this time, though in common with many other people we were the recipients of considerable financial assistance from his special charity fund.

Mother found London a depressing place to live and wanted to be near her circle of friends in Oxford, which included Jack, his brother "Warnie," and such people as Kay and Austin Farrer. I think it is too simple and too supposititious to say that her only motive for moving was to be near Jack, but it was certainly a contributory factor.

Our short time in Headington, just outside Oxford, seemed to be the beginning of so much that could have been wonderful. Our home was visited frequently by good friends and was the scene of many lively intellectual debates. It was also during this time that the relationship between Jack and Mother began to redefine itself.

I think that Jack resisted the deep emotional attachment to my mother which he began to be aware of, largely because it was something which he mistakenly thought was alien to his nature. Their friendship on a platonic level was convenient and caused no ripples on the placid surface of his existence. However, he was forced not merely to inward awareness of his love for her, but also to public acknowledgement of it by the sudden realisation that he was about to lose her.

It almost seems cruel that her death was delayed long enough for him to grow to love her so completely that she filled his world as the greatest gift that God had ever given him, and then she died and left him alone in a place that her presence in his life had created for him.

What many of us discover in this outpouring of anguish is that we know exactly what he is talking about. Those of us who have walked this same path, or are walking it as we read this book, find that we are not, after all, as alone as we thought.

C. S. Lewis, the writer of so much that is so clear and so right, the thinker whose acuity of mind and clarity of expression enabled us to understand so much, this strong and determined Christian, he too fell headlong into the vortex of whirling thoughts and feelings and dizzily groped for support and guidance deep in the dark chasm of grief. How I wish that he had been blessed with just such a book as this. If we find no comfort in the world around us, and no solace when we cry to God, if it does nothing else for us, at least this book will help us to face our grief, and to "misunderstand a little less completely."

For further reading, I recommend *Jack: C. S. Lewis and His Times* by George Sayer (Harper & Row, 1988; Crossway Books) as the best available biography of C. S. Lewis; Lyle Dorsett's biography of my mother, *And God Came In* (Macmillan, 1983); and also, somewhat immodestly perhaps, for an inside viewpoint of our family life, my own book, *Lenten Lands* (Macmillan, 1988; HarperSanFrancisco, 1994).

Douglas H. Gresham

A
GRIEF
OBSERVED

CHAPTER ONE

No one ever told me that grief felt so like fear. I am not afraid, but the sensation is like being afraid. The same fluttering in the stomach, the same restlessness, the yawning. I keep on swallowing.

At other times it feels like being mildly drunk, or concussed. There is a sort of invisible blanket between the world and me. I find it hard to take in what anyone says. Or perhaps, hard to want to take it in. It is so uninteresting. Yet I want the others to be about me. I dread the moments when the house is empty. If only they would talk to one another and not to me.

There are moments, most unexpectedly, when something inside me tries to assure me that I don't really mind so much, not so very much, after all. Love is not the whole of a man's life. I was happy before I ever met H. I've plenty of what are called 'resources.' People get over these things. Come, I shan't do so badly. One is ashamed to listen to this voice but it seems for a little to be making out a good case. Then comes a sudden jab of red-hot memory and all this 'commonsense' vanishes like an ant in the mouth of a furnace.

On the rebound one passes into tears and pathos. Maudlin tears. I almost prefer the moments of agony. These are at least clean and honest. But the bath of self-pity, the wallow, the loathsome sticky-sweet pleasure of indulging it—that disgusts me. And even while I'm doing it I know it leads me to misrepresent H. herself. Give that mood its head and in a few minutes I shall have substituted for the real woman a mere doll to be blubbered over. Thank God the memory of her is still too strong (will it always be too strong?) to let me get away with it.

For H. wasn't like that at all. Her mind was lithe and quick and muscular as a leopard. Passion, tenderness, and pain were all equally unable to disarm it. It scented the first whiff of cant or slush; then sprang, and knocked you over before you knew what was happening. How many bubbles of mine she pricked! I soon learned not to talk rot to her unless I did it for the sheer pleasure—and there's another red-hot jab—of being exposed and laughed at. I was never less silly than as H.'s lover.

And no one ever told me about the laziness of grief. Except at my job—where the machine seems to run on much as usual—I loathe the slightest effort. Not only writing but even reading a letter is too much. Even shaving. What does it matter now whether my cheek is rough or smooth? They say an unhappy man wants distractions—something to take him out of himself. Only as a dog-tired man wants an extra blanket on a cold night; he'd rather lie there shivering than get up and find one. It's easy to see why the lonely become untidy, finally, dirty and disgusting.

Meanwhile, where is God? This is one of the most disquieting symptoms. When you are happy, so happy that you have no sense of needing Him, so happy that you are tempted to feel His claims upon you as an interruption, if you remember yourself and turn to Him with gratitude and praise, you will be—or so it feels—welcomed with open arms. But go to Him when your need is desperate, when all other help is vain, and what do you find? A door slammed in your face, and a sound of bolting and double bolting on the inside. After that, silence. You may as well turn away. The longer you wait, the more emphatic the silence will become. There are no lights in the windows. It might be an empty house. Was it ever inhabited? It seemed so once. And that seeming was as strong as this. What can this mean? Why is He so present a commander in our time of prosperity and so very absent a help in time of trouble?

I tried to put some of these thoughts to C. this afternoon. He reminded me that the same thing seems to have happened to Christ: 'Why hast thou forsaken me?' I know. Does that make it easier to understand?

Not that I am (I think) in much danger of ceasing to believe in God. The real danger is of coming to believe such dreadful things about Him. The conclusion I dread is not 'So there's no God after all,' but 'So this is what God's really like. Deceive yourself no longer.'

Our elders submitted and said, 'Thy will be done.' How often had bitter resentment been stifled through sheer terror and an act of love—yes, in every sense, an act—put on to hide the operation?

Of course it's easy enough to say that God seems absent at our greatest need because He *is* absent—non-existent. But then why does He seem so present when, to put it quite frankly, we don't ask for Him?

One thing, however, marriage has done for me. I can never again believe that religion is manufactured out of our unconscious, starved desires and is a substitute for sex. For those few years H. and I feasted on love, every mode of it—solemn and merry, romantic and realistic, sometimes as dramatic as a thunderstorm, sometimes as comfortable and unemphatic as putting on your soft slippers. No cranny of heart or body remained unsatisfied. If God were a substitute for love we ought to have lost all interest in Him. Who'd bother about substitutes when he has the thing itself? But that isn't what happens. We both knew we wanted something besides one another—quite a different kind of something, a quite different kind of want. You might as well say that when lovers have one another they will never want to read, or eat—or breathe.

After the death of a friend, years ago, I had for some time a most vivid feeling of certainty about his continued life; even his enhanced life. I have begged to be given even one hundredth part of the same assurance about H. There is no answer. Only the locked door, the iron curtain, the vacuum, absolute zero. 'Them as asks don't get.' I was a fool to ask. For now, even if that assurance came I should distrust it. I should think it a self-hypnosis induced by my own prayers.

At any rate I must keep clear of the spiritualists. I promised H. I would. She knew something of those circles.

Keeping promises to the dead, or to anyone else, is very well. But I begin to see that 'respect for the wishes of the dead' is a trap. Yesterday I stopped myself only in time from saying about some trifle

'H. wouldn't have liked that.' This is unfair to the others. I should soon be using 'what H. would have liked' as an instrument of domestic tyranny, with her supposed likings becoming a thinner and thinner disguise for my own.

I cannot talk to the children about her. The moment I try, there appears on their faces neither grief, nor love, nor fear, nor pity, but the most fatal of all non-conductors, embarrassment. They look as if I were committing an indecency. They are longing for me to stop. I felt just the same after my own mother's death when my father mentioned her. I can't blame them. It's the way boys are.

I sometimes think that shame, mere awkward, senseless shame, does as much towards preventing good acts and straightforward happiness as any of our vices can do. And not only in boyhood.

Or are the boys right? What would H. herself think of this terrible little notebook to which I come back and back? Are these jottings morbid? I once read the sentence 'I lay awake all night with toothache, thinking about toothache and about lying awake.' That's true to life. Part of every misery is, so to speak, the misery's shadow or reflection: the fact that you don't merely suffer but have to keep on thinking about the fact that you suffer. I not only live each endless day in grief, but live each day thinking about living each day in grief. Do these notes merely aggravate that side of it? Merely confirm the monotonous, tread-mill march of the mind round one subject? But what am I to do? I must have some drug, and reading isn't a strong enough drug now. By writing it all down (all?—no: one thought in a hundred) I believe I get a little outside it. That's how I'd defend it to H. But ten to one she'd see a hole in the defence.

It isn't only the boys either. An odd byproduct of my loss is that I'm aware of being an embarrassment to everyone I meet. At work, at the club, in the street, I see people, as they approach me, trying to make up their minds whether they'll 'say something about it' or not. I hate it if they do, and if they don't. Some funk it altogether. R. has been avoiding me for a week. I like best the well brought-up young men, almost boys, who walk up to me as if I were a dentist, turn very red, get it over, and then edge away to the bar as quickly as they decently can. Perhaps the bereaved ought to be isolated in special settlements like lepers.

To some I'm worse than an embarrassment. I am a death's head. Whenever I meet a happily married pair I can feel them both thinking, 'One or other of us must some day be as he is now.'

At first I was very afraid of going to places where H. and I had been happy—our favourite pub, our favourite wood. But I decided to do it at once—like sending a pilot up again as soon as possible after he's had a crash. Unexpectedly, it makes no difference. Her absence is no more emphatic in those places than anywhere else. It's not local at all. I suppose that if one were forbidden all salt one wouldn't notice it much more in any one food than in another. Eating in general would be different, every day, at every meal. It is like that. The act of living is different all through. Her absence is like the sky, spread over everything.

But no, that is not quite accurate. There is one place where her absence comes locally home to me, and it is a place I can't avoid. I mean my own body. It had such a different importance while it was the body of H.'s lover. Now it's like an empty house. But don't let me deceive myself. This body would become important to me again, and pretty quickly, if I thought there was anything wrong with it.

Cancer, and cancer, and cancer. My mother, my father, my wife. I wonder who is next in the queue.

Yet H. herself, dying of it, and well knowing the fact, said that she had lost a great deal of her old horror at it. When the reality came, the name and the idea were in some degree disarmed. And up to a point I very nearly understood. This is important. One never meets just Cancer, or War, or Unhappiness (or Happiness). One only meets each hour or moment that comes. All manner of ups and downs. Many bad spots in our best times, many good ones in our worst. One never gets the total impact of what we call 'the thing itself.' But we call it wrongly. The thing itself is simply all these ups and downs: the rest is a name or an idea.

It is incredible how much happiness, even how much gaiety, we sometimes had together after all hope was gone. How long, how tranquilly, how nourishingly, we talked together that last night!

And yet, not quite together. There's a limit to the 'one flesh.' You can't really share someone else's weakness, or fear or pain. What you feel may be bad. It might conceivably be as bad as what the other felt, though I should distrust anyone who claimed that it was. But it would still be quite different. When I speak of fear, I mean the merely animal fear, the recoil of the organism from its destruction; the smothery feeling; the sense of being a rat in a trap. It can't be transferred. The mind can sympathize; the body, less. In one way the bodies of lovers can do it least. All their love passages have trained them to have, not identical, but complementary, correlative, even opposite, feelings about one another.

We both knew this. I had my miseries, not hers; she had hers, not mine. The end of hers would be the coming-of-age of mine. We were setting out on different roads. This cold truth, this terrible traffic-regulation ('You, Madam, to the right—you, Sir, to the left') is just the beginning of the separation which is death itself.

And this separation, I suppose, waits for all. I have been thinking of H. and myself as peculiarly unfortunate in being torn apart. But presumably all lovers are. She once said to me, 'Even if we both died at exactly the same moment, as we lie here side by side, it would be just as much a separation as the one you're so afraid of.' Of course she didn't *know,* any more than I do. But she was near death; near enough to make a good shot. She used to quote 'Alone into the Alone.' She said it felt like that. And how immensely improbable that it should be otherwise! Time and space and body were the very things that brought us together; the telephone wires by which we communicated. Cut one off, or cut both off simultaneously. Either way, mustn't the conversation stop?

Unless you assume that some other means of communication—utterly different, yet doing the same work—would be immediately substituted. But then, what conceivable point could there be in severing the old ones? Is God a clown who whips away your bowl of soup one moment in order, next moment, to replace it with another bowl of the same soup? Even nature isn't such a clown as that. She never plays exactly the same tune twice.

It is hard to have patience with people who say, 'There is no death' or 'Death doesn't matter.' There is death. And whatever is matters. And whatever happens has consequences, and it and they are irrevocable and irreversible. You might as well say that birth doesn't matter. I look up at the night sky. Is anything more certain than that in all those vast times and spaces, if I were allowed to search them, I should nowhere find her face, her voice, her touch? She died. She is dead. Is the word so difficult to learn?

I have no photograph of her that's any good. I cannot even see her face distinctly in my imagination. Yet the odd face of some stranger seen in a crowd this morning may come before me in vivid perfection the moment I close my eyes tonight. No doubt, the explanation is simple enough. We have seen the faces of those we know best so variously, from so many angles, in so many lights, with so many expressions—waking, sleeping, laughing, crying, eating, talking, thinking—that all the impressions crowd into our memory together and cancel out into a mere blur. But her voice is still vivid. The remembered voice—that can turn me at any moment to a whimpering child.

CHAPTER TWO

For the first time I have looked back and read these notes. They appall me. From the way I've been talking anyone would think that H.'s death mattered chiefly for its effect on myself. Her point of view seems to have dropped out of sight. Have I forgotten the moment of bitterness when she cried out, 'And there was so much to live for'? Happiness had not come to her early in life. A thousand years of it would not have made her *blasée.* Her palate for all the joys of sense and intellect and spirit was fresh and unspoiled. Nothing would have been wasted on her. She liked more things and liked them more than anyone I have known. A noble hunger, long unsatisfied, met at last its proper food, and almost instantly the food was snatched away. Fate (or whatever it is) delights to produce a great capacity and then frustrate it. Beethoven went deaf. By our standards a mean joke; the monkey trick of a spiteful imbecile.

I must think more about H. and less about myself.

Yes, that sounds very well. But there's a snag. I am thinking about her nearly always. Thinking of the H. facts—real words, looks, laughs, and actions of hers. But it is my own mind that selects and groups them. Already, less than a month after her death, I can feel the slow, insidious beginning of a process that will make the H. I think of into a more and more imaginary woman. Founded on fact, no doubt. I shall put in nothing fictitious (or I hope I shan't). But won't the composition inevitably become more and more my own? The reality is no longer there to check me, to pull me up short, as the real H. so often did, so unexpectedly, by being so thoroughly herself and not me.

The most precious gift that marriage gave me was this constant impact of something very close and intimate yet all the time unmistakably other, resistant—in a word, real. Is all that work to be undone? Is what I shall still call H. to sink back horribly into being not much more than one of my old bachelor pipe-dreams? Oh my dear, my dear, come back for one moment and drive that miserable phantom away. Oh God, God, why did you take such trouble to force this creature out of its shell if it is now doomed to crawl back—to be sucked back—into it?

Today I had to meet a man I haven't seen for ten years. And all that time I had thought I was remembering him well—how he looked and spoke and the sort of things he said. The first five minutes of the real man shattered the image completely. Not that he had changed. On the contrary. I kept on thinking, 'Yes, of course, of course. I'd forgotten that he thought that—or disliked this, or knew so-and-so—or jerked his head back that way.' I had known all these things once and I recognized them the moment I met them again. But they had all faded out of my mental picture of him, and when they were all replaced by his actual presence the total effect was quite astonishingly different from the image I had carried about with me for those ten years. How can I hope that this will not happen to my memory of H.? That it is not happening already? Slowly, quietly, like snow-flakes—like the small flakes that come when it is going to snow all night—little flakes of me, my impressions, my selections, are settling down on the image of her. The real shape will be quite hidden in the end. Ten minutes—ten seconds—of the real H. would correct all this. And yet, even if those ten seconds were allowed me, one second later the little flakes would begin to fall again. The rough, sharp, cleansing tang of her otherness is gone.

What pitiable cant to say, 'She will live forever in my memory!' *Live?* That is exactly what she won't do. You might as well think like the old Egyptians that you can keep the dead by embalming them. Will nothing persuade us that they are gone? What's left? A corpse, a memory, and (in some versions) a ghost. All mockeries or horrors. Three more ways of spelling the word *dead.* It was H. I loved. As if I wanted to fall in love with my memory of her, an image in my own mind! It would be a sort of incest.

I remember being rather horrified one summer morning long ago when a burly, cheerful labouring man, carrying a hoe and a watering pot came into our churchyard and, as he pulled the gate behind him, shouted over his shoulder to two friends, 'See you later, I'm just going to visit Mum.' He meant he was going to weed and water and generally tidy up her grave. It horrified me because this mode of sentiment, all this churchyard stuff, was and is simply hateful, even inconceivable, to me. But in the light of my recent thoughts I am beginning to wonder whether, if one could take that man's line (I can't), there isn't a good deal to be said for it. A six-by-three-foot flower-bed had become Mum. That was his symbol for her, his link with her. Caring for it was visiting her. May this not be in one way better than preserving and caressing an image in one's own memory? The grave and the image are equally links with the irrecoverable and symbols for the unimaginable. But the image has the added disadvantage that it will do whatever you want. It will smile or frown, be tender, gay, ribald, or argumentative just as your mood demands. It is a puppet of which you hold the strings. Not yet of course. The reality is still too fresh; genuine and wholly involuntary memories can still, thank God, at any moment rush in and tear the strings out of my hands. But the fatal obedience of the image, its insipid dependence on me, is bound to increase. The flower-bed on the other hand is an obstinate, resistant, often intractable bit of reality, just as Mum in her lifetime doubtless was. As H. was.

Or as H. is. Can I honestly say that I believe she now is anything? The vast majority of the people I meet, say, at work, would certainly think she is not. Though naturally they wouldn't press the point on me. Not just now anyway. What do I really think? I have always been able to pray for the other dead, and I still do, with some confidence. But when I try to pray for H., I halt. Bewilderment and amazement come over me. I have a ghastly sense of unreality, of speaking into a vacuum about a nonentity.

The reason for the difference is only too plain. You never know how much you really believe anything until its truth or falsehood becomes a matter of life and death to you. It is easy to say you believe a rope to be strong and sound as long as you are merely using it to cord a box. But suppose you had to hang by that rope over a precipice. Wouldn't you then first discover how much you really trusted it? The same with people. For years I would have said that I had perfect confidence in B.R. Then came the moment when I had to decide whether I would or would not trust him with a really important secret. That threw quite a new light on what I called my 'confidence' in him. I discovered that there was no such thing. Only a real risk tests the reality of a belief. Apparently the faith—I thought it faith—which enables me to pray for the other dead has seemed strong only because I have never really cared, not desperately, whether they existed or not. Yet I thought I did.

But there are other difficulties. 'Where is she now?' That is, *in what place* is she *at the present time?* But if H. is not a body—and the body I loved is certainly no longer she—she is in no place at all. And 'the present time' is a date or point in our time series. It is as if she were on a journey without me and I said, looking at my watch, 'I wonder is she at

Euston now.' But unless she is proceeding at sixty seconds a minute along this same timeline that all we living people travel by, what does *now* mean? If the dead are not in time, or not in our sort of time, is there any clear difference, when we speak of them, between *was* and *is* and *will be?*

Kind people have said to me, 'She is with God.' In one sense that is most certain. She is, like God, incomprehensible and unimaginable.

But I find that this question, however important it may be in itself, is not after all very important in relation to grief. Suppose that the earthly lives she and I shared for a few years are in reality only the basis for, or prelude to, or earthly appearance of, two unimaginable, supercosmic, eternal somethings. Those somethings could be pictured as spheres or globes. Where the plane of Nature cuts through them—that is, in earthly life—they appear as two circles (circles are slices of spheres). Two circles that touched. But those two circles, above all the point at which they touched, are the very thing I am mourning for, homesick for, famished for. You tell me, 'she goes on.' But my heart and body are crying out, come back, come back. Be a circle, touching my circle on the plane of Nature. But I know this is impossible. I know that the thing I want is exactly the thing I can never get. The old life, the jokes, the drinks, the arguments, the lovemaking, the tiny, heartbreaking commonplace. On any view whatever, to say, 'H. is dead,' is to say, 'All that is gone.' It is a part of the past. And the past is the past and that is what time means, and time itself is one more name for death, and Heaven itself is a state where 'the former things have passed away.'

Talk to me about the truth of religion and I'll listen gladly. Talk to me about the duty of religion and I'll listen submissively. But don't come talking to me about the consolations of religion or I shall suspect that you don't understand.

Unless, of course, you can literally believe all that stuff about family reunions 'on the further shore,' pictured in entirely earthly terms. But that is all unscriptural, all out of bad hymns and lithographs. There's not a word of it in the Bible. And it rings false. We *know* it couldn't be like that. Reality never repeats. The exact same thing is never taken away and given back. How well the spiritualists bait their hook! 'Things on this side are not so different after all.' There are cigars in Heaven. For that is what we should all like. The happy past restored.

And that, just that, is what I cry out for, with mad, midnight endearments and entreaties spoken into the empty air.

And poor C. quotes to me, 'Do not mourn like those that have no hope.' It astonishes me, the way we are invited to apply to ourselves words so obviously addressed to our betters. What St. Paul says can comfort only those who love God better than the dead, and the dead better than themselves. If a mother is mourning not for what she has lost but for what her dead child has lost, it is a comfort to believe that the child has not lost the end for which it was created. And it is a comfort to believe that she herself, in losing her chief or only natural happiness, has not lost a greater thing, that she may still hope to 'glorify God and enjoy Him forever.' A comfort to the God-aimed, eternal spirit within her. But not to her motherhood. The specifically maternal happiness must be written off. Never, in any place or time, will she have her son on her knees, or bathe him, or tell him a story, or plan for his future, or see her grandchild.

They tell me H. is happy now, they tell me she is at peace. What makes them so sure of this? I don't mean that I fear the worst of all. Nearly her last words were, 'I am at peace with God.' She had not always been. And she never lied. And she wasn't easily deceived, least of all, in her own favour. I don't mean that. But why are they so sure that all anguish ends with death? More than half the Christian world, and millions in the East, believe otherwise. How do they know she is 'at rest?' Why should the separation (if nothing else) which so agonizes the lover who is left behind be painless to the lover who departs?

'Because she is in God's hands.' But if so, she was in God's hands all the time, and I have seen what they did to her here. Do they suddenly become gentler to us the moment we are out of the body? And if so, why? If God's goodness is inconsistent with hurting us, then either God is not good or there is no God: for in the only life we know He hurts us beyond our worst fears and beyond all we can imagine. If it is consistent with hurting us, then He may hurt us after death as unendurably as before it.

Sometimes it is hard not to say, 'God forgive God.' Sometimes it is hard to say so much. But if our faith is true, He didn't. He crucified Him.

Come, what do we gain by evasions? We are under the harrow and can't escape. Reality, looked at steadily, is unbearable. And how or why did such a reality blossom (or fester) here and there into the terrible phenomenon called consciousness? Why did it produce things like us who can see it and, seeing it, recoil in loathing? Who (stranger still) want to see it and take pains to find it out, even when no need compels them and even though the sight of it makes an incurable ulcer in their hearts? People like H. herself, who would have truth at any price.

If H. 'is not,' then she never was. I mistook a cloud of atoms for a person. There aren't, and never were, any people. Death only reveals the vacuity that was always there. What we call the living are simply those who have not yet been unmasked. All equally bankrupt, but some not yet declared.

But this must be nonsense; vacuity revealed to whom? Bankruptcy declared to whom? To other boxes of fireworks or clouds of atoms. I will never believe—more strictly I can't believe—that one set of physical events could be, or make, a mistake about other sets.

No, my real fear is not of materialism. If it were true, we—or what we mistake for 'we'—could get out, get from under the harrow. An overdose of sleeping pills would do it. I am more afraid that we are really rats in a trap. Or, worse still, rats in a laboratory. Someone said, I believe, 'God always geometrizes.' Supposing the truth were 'God always vivisects'?

Sooner or later I must face the question in plain language. What reason have we, except our own desperate wishes, to believe that God is, by any standard we can conceive, 'good'? Doesn't all the *prima facie* evidence suggest exactly the opposite? What have we to set against it?

We set Christ against it. But how if He were mistaken? Almost His last words may have a perfectly clear meaning. He had found that the Being He called Father was horribly and infinitely different from what He had supposed. The trap, so long and carefully prepared and so subtly baited, was at last sprung, on the cross. The vile practical joke had succeeded.

What chokes every prayer and every hope is the memory of all the prayers H. and I offered and all the false hopes we had. Not hopes raised merely by our own wishful thinking, hopes encouraged, even forced upon us, by false diagnoses, by X-ray photographs, by strange remissions, by one temporary recovery that might have ranked as a miracle. Step by step we were 'led up the garden path.' Time after time, when He seemed most gracious He was really preparing the next torture.

I wrote that last night. It was a yell rather than a thought. Let me try it over again. Is it rational to believe in a bad God? Anyway, in a God so bad as all that? The Cosmic Sadist, the spiteful imbecile?

I think it is, if nothing else, too anthropomorphic. When you come to think of it, it is far more anthropomorphic than picturing Him as a grave old king with a long beard. That image is a Jungian archetype. It links God with all the wise old kings in the fairy-tales, with prophets, sages, magicians. Though it is (formally) the picture of a man, it suggests something more than humanity. At the very least it gets in the idea of something older than yourself, something that knows more, something you can't fathom. It preserves mystery. Therefore room for hope. Therefore room for a dread or awe that needn't be mere fear of mischief from a spiteful potentate. But the picture I was building up last night is simply the picture of a man like S.C.—who used to sit next to me at dinner and tell me what he'd been doing to the cats that afternoon. Now a being like S.C., however magnified, couldn't invent or create or govern anything. He would set traps and try to bait them. But he'd never have thought of baits like love, or laughter, or daffodils, or a frosty sunset. *He* make a universe? He couldn't make a joke, or a bow, or an apology, or a friend.

Or could one seriously introduce the idea of a bad God, as it were by the back door, through a sort of extreme Calvinism? You could say we are fallen and depraved. We are so depraved that our ideas of goodness count for nothing; or worse than nothing—the very fact that we think something good is presumptive evidence that it is really bad. Now God has in fact—our worst fears are true—all the characteristics we regard as bad: unreasonableness, vanity, vindictiveness, injustice, cruelty. But all these blacks (as they seem to us) are really whites. It's only our depravity that makes them look black to us.

And so what? This, for all practical (and speculative) purposes, sponges God off the slate. The word *good,* applied to Him, becomes meaningless: like abracadabra. We have no motive for obeying Him. Not even fear. It is true we have His threats and promises. But why should we believe them? If cruelty is from His point of view 'good,' telling lies may be 'good' too. Even if they are true, what then? If His ideas of good are so very different from ours, what He calls Heaven might well be what we should call Hell, and vice-versa. Finally, if reality at its very root is so meaningless to us—or, putting it the other way round, if we are such total imbeciles—what is the point of trying to think either about God or about anything else? This knot comes undone when you try to pull it tight.

Why do I make room in my mind for such filth and nonsense? Do I hope that if feeling disguises itself as thought I shall feel less? Aren't all these notes the senseless writhings of a man who won't accept the fact that there is nothing we can do with suffering except to suffer it? Who still thinks there is some device (if only he could find it) which will make pain not to be pain. It doesn't really matter whether you grip the arms of the dentist's chair or let your hands lie in your lap. The drill drills on.

And grief still feels like fear. Perhaps, more strictly, like suspense. Or like waiting; just hanging about waiting for something to happen. It gives life a permanently provisional feeling. It doesn't seem worth starting anything. I can't settle down. I yawn, I fidget, I smoke too much. Up till this I always had too little time. Now there is nothing but time. Almost pure time, empty successiveness.

One flesh. Or, if you prefer, one ship. The starboard engine has gone. I, the port engine, must chug along somehow till we make harbour. Or rather, till the journey ends. How can I assume a harbour? A lee shore, more likely, a black night, a deafening gale, breakers ahead—and any lights shown from the land probably being waved by wreckers. Such was H.'s landfall. Such was my mother's. I say their landfalls; not their arrivals.

CHAPTER THREE

It's not true that I'm always thinking of H. Work and conversation make that impossible. But the times when I'm not are perhaps my worst. For then, though I have forgotten the reason, there is spread over everything a vague sense of wrongness, of something amiss. Like in those dreams where nothing terrible occurs—nothing that would sound even remarkable if you told it at breakfast-time—but the atmosphere, the taste, of the whole thing is deadly. So with this. I see the rowan berries reddening and don't know for a moment why they, of all things, should be depressing. I hear a clock strike and some quality it always had before has gone out of the sound. What's wrong with the world to make it so flat, shabby, worn-out looking? Then I remember.

This is one of the things I'm afraid of. The agonies, the mad midnight moments, must, in the course of nature, die away. But what will follow? Just this apathy, this dead flatness? Will there come a time when I no longer ask why the world is like a mean street, because I shall take the squalor as normal? Does grief finally subside into boredom tinged by faint nausea?

Feelings, and feelings, and feelings. Let me try thinking instead. From the rational point of view, what new factor has H.'s death introduced into the problem of the universe? What grounds has it given me for doubting all that I believe? I knew already that these things, and worse, happened daily. I would have said that I had taken them into account. I had been warned—I had warned myself—not to reckon on worldly happiness. We were even promised sufferings. They were part of the programme. We were even told, 'Blessed are they that mourn,' and I accepted it. I've got nothing that I hadn't bargained for. Of course it is different when the thing happens to oneself, not to others, and in reality, not in imagination. Yes; but should it, for a sane man, make quite such a difference as this? No. And it wouldn't for a man whose faith had been real faith and whose concern for other people's sorrows had been real concern. The case is too plain. If my house has collapsed at one blow, that is because it was a house of cards. The faith which 'took these things into account' was not faith but imagination. The taking them into account was not real sympathy. If I had really cared, as I thought I did, about the sorrows of the world, I should not have been so overwhelmed when my own sorrow came. It has been an imaginary faith playing with innocuous counters labelled 'Illness,' 'Pain,' 'Death,' and 'Loneliness.' I thought I trusted the rope until it mattered to me whether it would bear me. Now it matters, and I find I didn't.

Bridge-players tell me that there must be some money on the game 'or else people won't take it seriously.' Apparently it's like that. Your bid—for God or no God, for a good God or the Cosmic Sadist, for eternal life or nonentity—will not be serious if nothing much is staked on it. And you will never discover how serious it was until the stakes are raised horribly high, until you find that you are playing not for counters or for sixpences but for every penny you have in the world. Nothing less will shake a man—or at any rate a man like me—out of his merely verbal thinking and his merely notional beliefs. He has to be knocked silly before he comes to his senses. Only torture will bring out the truth. Only under torture does he discover it himself.

And I must surely admit—H. would have forced me to admit in a few passes—that, if my house was a house of cards, the sooner it was knocked down the better. And only suffering could do it. But then the Cosmic Sadist and Eternal Vivisector becomes an unnecessary hypothesis.

Is this last note a sign that I'm incurable, that when reality smashes my dream to bits, I mope and snarl while the first shock lasts, and then patiently, idiotically, start putting it together again? And so always? However often the house of cards falls, shall I set about rebuilding it? Is that what I'm doing now?

Indeed it's likely enough that what I shall call, if it happens, a 'restoration of faith' will turn out to be only one more house of cards. And I shan't know whether it is or not until the next blow comes—when, say, fatal disease is diagnosed in my body too, or war breaks out, or I have ruined myself by some ghastly mistake in my work. But there are two questions here. In which sense may it be a house of cards? Because the things I am believing are only a dream, or because I only dream that I believe them?

As for the things themselves, why should the thoughts I had a week ago be any more trustworthy than the better thoughts I have now? I am surely, in general, a saner man than I was then. Why should the desperate imaginings of a man dazed—I said it was like being concussed—be especially reliable?

Because there was no wishful thinking in them? Because, being so horrible, they were therefore all the more likely to be true? But there are fear-fulfilment as well as wish-fulfilment dreams. And were they wholly distasteful? No. In a way I liked them. I am even aware of a slight reluctance to accept the opposite thoughts. All that stuff about the Cosmic

Sadist was not so much the expression of thought as of hatred. I was getting from it the only pleasure a man in anguish can get; the pleasure of hitting back. It was really just Billingsgate—mere abuse; 'telling God what I thought of Him.' And of course, as in all abusive language, 'what I thought' didn't mean what I thought true. Only what I thought would offend Him (and His worshippers) most. That sort of thing is never said without some pleasure. Gets it 'off your chest.' You feel better for a moment.

But the mood is no evidence. Of course the cat will growl and spit at the operator and bite him if she can. But the real question is whether he is a vet or a vivisector. Her bad language throws no light on it one way or the other.

And I can believe He is a vet when I think of my own suffering. It is harder when I think of hers. What is grief compared with physical pain? Whatever fools may say, the body can suffer twenty times more than the mind. The mind has always some power of evasion. At worst, the unbearable thought only comes back and back, but the physical pain can be absolutely continuous. Grief is like a bomber circling round and dropping its bombs each time the circle brings it overhead; physical pain is like the steady barrage on a trench in World War One, hours of it with no let-up for a moment. Thought is never static; pain often is.

What sort of a lover am I to think so much about my affliction and so much less about hers? Even the insane call, 'Come back,' is all for my own sake. I never even raised the question whether such a return, if it were possible, would be good for her. I want her back as an ingredient in the restoration of *my* past. Could I have wished her anything worse? Having got once through death, to come back and then, at some later date, have all her dying to do over again? They call Stephen the first martyr. Hadn't Lazarus the rawer deal?

I begin to see. My love for H. was of much the same quality as my faith in God. I won't exaggerate, though. Whether there was anything but imagination in the faith, or anything but egoism in the love, God knows. I don't. There may have been a little more; especially in my love for H. But neither was the thing I thought it was. A good deal of the card-castle about both.

What does it matter how this grief of mine evolves or what I do with it? What does it matter how I remember her or whether I remember her at all? None of these alternatives will either ease or aggravate her past anguish.

Her past anguish. How do I know that all her anguish is past? I never believed before—I thought it immensely improbable—that the faithfulest soul could leap straight into perfection and peace the moment death has rattled in the throat. It would be wishful thinking with a vengeance to take up that belief now. H. was a splendid thing; a soul straight, bright, and tempered like a sword. But not a perfected saint. A sinful woman married to a sinful man; two of God's patients, not yet cured. I know there are not only tears to be dried but stains to be scoured. The sword will be made even brighter.

But oh God, tenderly, tenderly. Already, month by month and week by week you broke her body on the wheel whilst she still wore it. Is it not yet enough?

The terrible thing is that a perfectly good God is in

this matter hardly less formidable than a Cosmic Sadist. The more we believe that God hurts only to heal, the less we can believe that there is any use in begging for tenderness. A cruel man might be bribed—might grow tired of his vile sport—might have a temporary fit of mercy, as alcoholics have fits of sobriety. But suppose that what you are up against is a surgeon whose intentions are wholly good. The kinder and more conscientious he is, the more inexorably he will go on cutting. If he yielded to your entreaties, if he stopped before the operation was complete, all the pain up to that point would have been useless. But is it credible that such extremities of torture should be necessary for us? Well, take your choice. The tortures occur. If they are unnecessary, then there is no God or a bad one. If there is a good God, then these tortures are necessary. For no even moderately good Being could possibly inflict or permit them if they weren't.

Either way, we're for it.

What do people mean when they say, 'I am not afraid of God because I know He is good'? Have they never even been to a dentist?

Yet this is unendurable. And then one babbles—'If only I could bear it, or the worst of it, or any of it, instead of her.' But one can't tell how serious that bid is, for nothing is staked on it. If it suddenly became a real possibility, then, for the first time, we should discover how seriously we had meant it. But is it ever allowed?

It was allowed to One, we are told, and I find I can now believe again, that He has done vicariously whatever can be so done. He replies to our babble, 'You cannot and you dare not. I could and dared.'

Something quite unexpected has happened. It came this morning early. For various reasons, not in themselves at all mysterious, my heart was lighter than it had been for many weeks. For one thing, I suppose I am recovering physically from a good deal of mere exhaustion. And I'd had a very tiring but very healthy twelve hours the day before, and a sounder night's sleep; and after ten days of low-hung grey skies and motionless warm dampness, the sun was shining and there was a light breeze. And suddenly at the very moment when, so far, I mourned H. least, I remembered her best. Indeed it was something (almost) better than memory; an instantaneous, unanswerable impression. To say it was like a meeting would be going too far. Yet there was that in it which tempts one to use those words. It was as if the lifting of the sorrow removed a barrier.

Why has no one told me these things? How easily I might have misjudged another man in the same situation? I might have said, 'He's got over it. He's forgotten his wife,' when the truth was, 'He remembers her better *because* he has partly got over it.'

Such was the fact. And I believe I can make sense out of it. You can't see anything properly while your eyes are blurred with tears. You can't, in most things, get what you want if you want it too desperately: anyway, you can't get the best out of it. 'Now! Let's have a real good talk' reduces everyone to silence. 'I *must* get a good sleep tonight' ushers in hours of wakefulness. Delicious drinks are wasted on a really ravenous thirst. Is it similarly the very intensity of the longing that draws the iron curtain, that makes us feel we are staring into a vacuum when we think about our dead? 'Them as asks' (at any rate 'as asks too importunately') don't get. Perhaps can't.

And so, perhaps, with God. I have gradually been coming to feel that the door is no longer shut and bolted. Was it my own frantic need that slammed it in my face? The time when there is nothing at all in your soul except a cry for help may be just the time when God can't give it: you are like the drowning man who can't be helped because he clutches and grabs. Perhaps your own reiterated cries deafen you to the voice you hoped to hear.

On the other hand, 'Knock and it shall be opened.' But does knocking mean hammering and kicking the door like a maniac? And there's also 'To him that hath shall be given.' After all, you must have a capacity to receive, or even omnipotence can't give. Perhaps your own passion temporarily destroys the capacity.

For all sorts of mistakes are possible when you are dealing with Him. Long ago, before we were married, H. was haunted all one morning as she went about her work with the obscure sense of God (so to speak) 'at her elbow,' demanding her attention. And of course, not being a perfected saint, she had the feeling that it would be a question, as it usually is, of some unrepented sin or tedious duty. At last she gave in—I know how one puts it off—and faced Him. But the message was, 'I want to *give* you something' and instantly she entered into joy.

I think I am beginning to understand why grief feels like suspense. It comes from the frustration of so many impulses that had become habitual. Thought after thought, feeling after feeling, action after action, had H. for their object. Now their target is gone. I keep on through habit fitting an arrow to the string, then I remember and have to lay the bow down. So many roads lead thought to H. I set out on one of them. But now there's an impassable frontierpost across it. So many roads once; now so many *culs de sac.*

For a good wife contains so many persons in herself. What was H. not to me? She was my daughter and my mother, my pupil and my teacher, my subject and my sovereign; and always, holding all these in solution, my trusty comrade, friend, shipmate, fellow-soldier. My mistress; but at the same time all that any man friend (and I have good ones) has ever been to me. Perhaps more. If we had never fallen in love we should have none the less been always together, and created a scandal. That's what I meant when I once praised her for her 'masculine virtues.' But she soon put a stop to that by asking how I'd like to be praised for my feminine ones. It was a good *riposte,* dear. Yet there was something of the Amazon, something of Penthesileia and Camilla. And you, as well as I, were glad it should be there. You were glad I should recognize it.

Solomon calls his bride Sister. Could a woman be a complete wife unless, for a moment, in one particular mood, a man felt almost inclined to call her Brother?

'It was too perfect to last,' so I am tempted to say of our marriage. But it can be meant in two ways. It may be grimly pessimistic—as if God no sooner saw two of His creatures happy than He stopped it ('None of that here!'). As if He were like the Hostess at the sherry-party who separates two guests the moment they show signs of having got into a real conversation. But it could also mean 'This had reached its proper perfection. This had become what it had in it to be. Therefore of course it would not be prolonged.' As if God said, 'Good; you have mastered that exercise. I am very pleased with it. And now you are ready to go on to the next.' When you have learned to do quadratics and enjoy doing them you will not be set them much longer. The teacher moves you on.

For we did learn and achieve something. There is, hidden or flaunted, a sword between the sexes till an entire marriage reconciles them. It is arrogance in us to call frankness, fairness, and chivalry 'masculine' when we see them in a woman; it is arrogance in them to describe a man's sensitiveness or tact or tenderness as 'feminine.' But also what poor, warped fragments of humanity most mere men and mere women must be to make the implications of that arrogance plausible. Marriage heals this. Jointly the two become fully human. 'In the image of God created He *them.*' Thus, by a paradox, this carnival of sexuality leads us out beyond our sexes.

And then one or other dies. And we think of this as love cut short; like a dance stopped in mid-career or a flower with its head unluckily snapped off—something truncated and therefore, lacking its due shape. I wonder. If, as I can't help suspecting, the dead also feel the pains of separation (and this may be one of their purgatorial sufferings), then for both lovers, and for all pairs of lovers without exception, bereavement is a universal and integral part of our experience of love. It follows marriage as normally as marriage follows courtship or as autumn follows summer. It is not a truncation of the process but one of its phases; not the interruption of the dance, but the next figure. We are 'taken out of ourselves' by the loved one while she is here. Then comes the tragic figure of the dance in which we must learn to be still taken out of ourselves though the bodily presence is withdrawn, to love the very Her, and not fall back to loving our past, or our memory, or our sorrow, or our relief from sorrow, or our own love.

Looking back, I see that only a very little time ago I was greatly concerned about my memory of H. and how false it might become. For some reason—the merciful good sense of God is the only one I can think of—I have stopped bothering about that. And the remarkable thing is that since I stopped bothering about it, she seems to meet me everywhere. *Meet* is far too strong a word. I don't mean anything remotely like an apparition or a voice. I don't mean even any strikingly emotional experience at any particular moment. Rather, a sort of unobtrusive but massive sense that she is, just as much as ever, a fact to be taken into account.

'To be taken into account' is perhaps an unfortunate way of putting it. It sounds as if she were rather a battle-axe. How can I put it better? Would 'momentously real' or 'obstinately real' do? It is as if the experience said to me, 'You are, as it happens, extremely glad that H. is still a fact. But remember she would be equally a fact whether you liked it or not. Your preferences have not been considered.'

How far have I got? Just as far, I think, as a widower of another sort who would stop, leaning on his spade, and say in answer to our inquiry, 'Thank'ee. Mustn't grumble. I do miss her something dreadful. But they say these things are sent to try us.' We have come to the same point; he with his spade, and I, who am not now much good at digging, with my own instrument. But of course one must take 'sent to try us' the right way. God has not been trying an experiment on my faith or love in order to find out their quality. He knew it already. It was I who didn't. In this trial He makes us occupy the dock, the witness box, and the bench all at once. He always knew that my temple was a house of cards. His only way of making me realize the fact was to knock it down.

Getting over it so soon? But the words are ambiguous. To say the patient is getting over it after an operation for appendicitis is one thing; after he's had his leg off it is quite another. After that operation either the wounded stump heals or the man dies. If it heals, the fierce, continuous pain will stop. Presently he'll get back his strength and be able to stump about on his wooden leg. He has 'got over it.' But he will probably have recurrent pains in the stump all his life, and perhaps pretty bad ones; and he will always be a one-legged man. There will be hardly any moment when he forgets it. Bathing, dressing, sitting down and getting up again, even lying in bed, will all be different. His whole way of life will be changed. All sorts of pleasures and activities that he once took for granted will have to be simply written off. Duties too. At present I am learning to get about on crutches. Perhaps I shall presently be given a wooden leg. But I shall never be a biped again.

Still, there's no denying that in some sense I 'feel better,' and with that comes at once a sort of shame, and a feeling that one is under a sort of obligation to cherish and foment and prolong one's unhappiness. I've read about that in books, but I never dreamed I should feel it myself. I am sure H. wouldn't approve of it. She'd tell me not to be a fool. So I'm pretty certain, would God. What is behind it?

Partly, no doubt, vanity. We want to prove to ourselves that we are lovers on the grand scale, tragic heroes; not just ordinary privates in the huge army of the bereaved, slogging along and making the best of a bad job. But that's not the whole of the explanation.

I think there is also a confusion. We don't really want grief, in its first agonies, to be prolonged: nobody could. But we want something else of which grief is a frequent symptom, and then we confuse the symptom with the thing itself. I wrote the other night that bereavement is not the truncation of married love but one of its regular phases—like the honeymoon. What we want is to live our marriage well and faithfully through that phase too. If it hurts (and it certainly will) we accept the pains as a necessary part of this phase. We don't want to escape them at the price of desertion or divorce. Killing the dead a second time. We were one flesh. Now that it has been cut in two, we don't want to pretend that it is whole and complete. We will be still married, still in love. Therefore we shall still ache. But we are not at all—if we understand ourselves—seeking the aches for their own sake. The less of them the better, so long as the marriage is preserved. And the more joy there can be in the marriage between dead and living, the better.

The better in every way. For, as I have discovered, passionate grief does not link us with the dead but cuts us off from them. This become clearer and clearer. It is just at those moments when I feel least sorrow—getting into my morning bath is usually one of them—that H. rushes upon my mind in her full reality, her otherness. Not, as in my worst moments, all foreshortened and patheticized and solemnized by my miseries, but as she is in her own right. This is good and tonic.

I seem to remember—though I couldn't quote one at the moment—all sorts of ballads and folk-tales in which the dead tell us that our mourning does them some kind of wrong. They beg us to stop it. There may be far more depth in this than I thought. If so, our grandfathers' generation went very far astray. All that (sometimes lifelong) ritual of sorrow—visiting graves, keeping anniversaries, leaving the empty bedroom exactly as 'the departed' used to keep it, mentioning the dead either not at all or always in a special voice, or even (like Queen Victoria) having the dead man's clothes put out for dinner every evening—this was like mummification. It made the dead far more dead.

Or was that (unconsciously) its purpose? Something very primitive may be at work here. To keep the dead thoroughly dead, to make sure that they won't come sidling back among the living, is a main pre-occupation of the savage mind. At all costs make them 'stay put.' Certainly these rituals do in fact emphasize their deadness. Perhaps this result was not really so unwelcome, not always, as the ritualists believed.

But I've no business to judge them. All guesswork; I'd better keep my breath to cool my own porridge. For me at any rate the programme is plain. I will turn to her as often as possible in gladness. I will even salute her with a laugh. The less I mourn her the nearer I seem to her.

An admirable programme. Unfortunately it can't be carried out. Tonight all the hells of young grief have opened again; the mad words, the bitter resentment, the fluttering in the stomach, the nightmare unreality, the wallowed-in tears. For in grief nothing 'stays put.' One keeps on emerging from a phase, but it always recurs. Round and round. Everything repeats. Am I going in circles, or dare I hope I am on a spiral?

But if a spiral, am I going up or down it?

How often—will it be for always?—how often will the vast emptiness astonish me like a complete novelty and make me say, 'I never realized my loss till this moment'? The same leg is cut off time after time. The first plunge of the knife into the flesh is felt again and again.

They say, 'The coward dies many times'; so does the beloved. Didn't the eagle find a fresh liver to tear in Prometheus every time it dined?

CHAPTER FOUR

This is the fourth—and the last—empty MS. book I can find in the house; at least nearly empty, for there are some pages of very ancient arithmetic at the end by J. I resolve to let this limit my jottings. I *will not* start buying books for the purpose. In so far as this record was a defence against total collapse, a safety-valve, it has done some good. The other end I had in view turns out to have been based on a misunderstanding. I thought I could describe a *state;* make a map of sorrow. Sorrow, however, turns out to be not a state but a process. It needs not a map but a history, and if I don't stop writing that history at some quite arbitrary point, there's no reason why I should ever stop. There is something new to be chronicled every day. Grief is like a long valley, a winding valley where any bend may reveal a totally new landscape. As I've already noted, not every bend does. Sometimes the surprise is the opposite one; you are presented with exactly the same sort of country you thought you had left behind miles ago. That is when you wonder whether the valley isn't a circular trench. But it isn't. There are partial recurrences, but the sequence doesn't repeat.

Here, for instance, is a new phase, a new loss. I do all the walking I can, for I'd be a fool to go to bed not tired. Today I have been revisiting old haunts, taking one of the long rambles that made me so happy in my bachelor days. And this time the face of nature was not emptied of its beauty and the world didn't look (as I complained some days ago) like a mean street. On the contrary, every horizon, every stile or clump of trees, summoned me into a past kind of happiness, my pre-H. happiness. But the invitation seemed to me horrible. The happiness into which it invited me was insipid. I find that I don't want to go back again and be happy in *that* way. It frightens me to think that a mere going back should even be possible. For this fate would seem to me the worst of all, to reach a state in which my years of love and marriage should appear in retrospect a charming episode—like a holiday—that had briefly interrupted my interminable life and returned me to normal, unchanged. And then it would come to seem unreal—something so foreign to the usual texture of my history that I could almost believe it had happened to someone else. Thus H. would die to me a second time; a worse bereavement than the first. Anything but that.

Did you ever know, dear, how much you took away with you when you left? You have stripped me even of my past, even of the things we never shared. I was wrong to say the stump was recovering from the pain of the amputation. I was deceived because it has so many ways to hurt me that I discover them only one by one.

Still, there are the two enormous gains—I know myself too well now to call them 'lasting.' Turned to God, my mind no longer meets that locked door; turned to H., it no longer meets that vacuum—nor all that fuss about my mental image of her. My jottings show something of the process, but not so much as I'd hoped. Perhaps both changes were really not observable. There was no sudden, striking, and emotional transition. Like the warming of a room or the coming of daylight. When you first notice them they have already been going on for some time.

The notes have been about myself, and about H., and about God. In that order. The order and the proportions exactly what they ought not to have been. And I see that I have nowhere fallen into that mode of thinking about either which we call praising them. Yet that would have been best for me. Praise is the mode of love which always has some element of joy in it. Praise in due order; of Him as the giver, of her as the gift. Don't we in praise somehow enjoy what we praise, however far we are from it? I must do more of this. I have lost the fruition I once had of H. And I am far, far away in the valley of my unlikeness, from the fruition which, if His mercies are infinite, I may some time have of God. But by praising I can still, in some degree, enjoy her, and already, in some degree, enjoy Him. Better than nothing.

But perhaps I lack the gift. I see I've described H. as being like a sword. That's true as far as it goes. But utterly inadequate by itself, and misleading. I ought to have balanced it. I ought to have said, 'But also like a garden. Like a nest of gardens, wall within wall, hedge within hedge, more secret, more full of fragrant and fertile life, the further you entered.'

And then, of her, and of every created thing I praise, I should say, 'In some way, in its unique way, like Him who made it.'

Thus up from the garden to the Gardener, from the sword to the Smith. To the life-giving Life and the Beauty that makes beautiful.

'She is in God's hands.' That gains a new energy when I think of her as a sword. Perhaps the earthly life I shared with her was only part of the tempering. Now perhaps He grasps the hilt; weighs the new weapon; makes lightnings with it in the air. 'A right Jerusalem blade.'

One moment last night can be described in similes; otherwise it won't go into language at all.

Imagine a man in total darkness. He thinks he is in a cellar or dungeon. Then there comes a sound. He thinks it might be a sound from far off—waves or wind-blown trees or cattle half a mile away. And if so, it proves he's not in a cellar, but free, in the open air. Or it may be a much smaller sound close at hand—a chuckle of laughter. And if so, there is a friend just beside him in the dark. Either way, a good, good sound. I'm not mad enough to take such an experience as evidence for anything. It is simply the leaping into imaginative activity of an idea which I would always have theoretically admitted—the idea that I, or any mortal at any time, may be utterly mistaken as to the situation he is really in.

Five senses; an incurably abstract intellect; a haphazardly selective memory; a set of preconceptions and assumptions so numerous that I can never examine more than a minority of them—never become even conscious of them all. How much of total reality can such an apparatus let through?

I will not, if I can help it, shin up either the feathery or the prickly tree. Two widely different convictions press more and more on my mind. One is that the Eternal Vet is even more inexorable and the possible operations even more painful than our severest imaginings can forbode. But the other, that 'all shall be well, and all shall be well, and all manner of thing shall be well.'

It doesn't matter that all the photographs of H. are bad. It doesn't matter—not much—if my memory of her is imperfect. Images, whether on paper or in the mind, are not important for themselves. Merely links. Take a parallel from an infinitely higher sphere. Tomorrow morning a priest will give me a little round, thin, cold, tasteless wafer. Is it a disadvantage—is it not in some ways an advantage—that it can't pretend the least *resemblance* to that with which it unites me?

I need Christ, not something that resembles Him. I want H., not something that is like her. A really good photograph might become in the end a snare, a horror, and an obstacle.

Images, I must suppose, have their use or they would not have been so popular. (It makes little difference whether they are pictures and statues outside the mind or imaginative constructions within it.) To me, however, their danger is more obvious. Images of the Holy easily become holy images—sacrosanct. My idea of God is not a divine idea. It has to be shattered time after time. He shatters it Himself. He is the great iconoclast. Could we not almost say that this shattering is one of the marks of His presence? The Incarnation is the supreme example; it leaves all previous ideas of the Messiah in ruins. And most are 'offended' by the iconoclasm; and blessed are those who are not. But the same thing happens in our private prayers.

All reality is iconoclastic. The earthly beloved, even in this life, incessantly triumphs over your mere idea of her. And you want her to; you want her with all her resistances, all her faults, all her unexpectedness. That is, in her foursquare and independent reality. And this, not any image or memory, is what we are to love still, after she is dead.

But 'this' is not now imaginable. In that respect H. and all the dead are like God. In that respect loving her has become, in its measure, like loving Him. In both cases I must stretch out the arms and hands of love—its eyes cannot here be used—to the reality, through—across—all the changeful phantasmagoria of my thoughts, passions, and imaginings. I mustn't sit down content with the phantasmagoria itself and worship that for Him, or love that for her.

Not my idea of God, but God. Not my idea of H., but H. Yes, and also not my idea of my neighbour, but my neighbour. For don't we often make this mistake as regards people who are still alive—who are with us in the same room? Talking and acting not to the man himself but to the picture—almost the *précis*—we've made of him in our own minds? And he has to depart from it pretty widely before we even notice the fact. In real life—that's one way it differs from novels—his words and acts are, if we observe closely, hardly ever quite 'in character,' that is, in what we call his character. There's always a card in his hand we didn't know about.

My reason for assuming that I do this to other people is the fact that so often I find them obviously doing it to me. We all think we've got one another taped.

And all this time I may, once more, be building with cards. And if I am He will once more knock the building flat. He will knock it down as often as proves necessary. Unless I have to be finally given up as hopeless, and left building pasteboard palaces in Hell forever; 'free among the dead.'

Am I, for instance, just sidling back to God because I know that if there's any road to H., it runs through Him? But then of course I know perfectly well that He can't be used as a road. If you're approaching Him not as the goal but as a road, not as the end but as a means, you're not really approaching Him at all. That's what was really wrong with all those popular pictures of happy reunions 'on the further shore'; not the simple-minded and very earthly images, but the fact that they make an End of what we can get only as a by-product of the true End.

Lord, are these your real terms? Can I meet H. again only if I learn to love you so much that I don't care whether I meet her or not? Consider, Lord, how it looks to us. What would anyone think of me if I said to the boys, 'No toffee now. But when you've grown up and don't really want toffee you shall have as much of it as you choose'?

If I knew that to be eternally divided from H. and eternally forgotten by her would add a greater joy and splendour to her being, of course I'd say, 'Fire ahead.' Just as if, on earth, I could have cured her cancer by never seeing her again, I'd have arranged never to see her again. I'd have had to. Any decent person would. But that's quite different. That's not the situation I'm in.

When I lay these questions before God I get no answer. But a rather special sort of 'No answer.' It is not the locked door. It is more like a silent, certainly not uncompassionate, gaze. As though He shook His head not in refusal but waiving the question. Like, 'Peace, child; you don't understand.'

Can a mortal ask questions which God finds unanswerable? Quite easily, I should think. All nonsense questions are unanswerable. How many hours are there in a mile? Is yellow square or round? Probably half the questions we ask—half our great theological and metaphysical problems—are like that.

And now that I come to think of it, there's no practical problem before me at all. I know the two great commandments, and I'd better get on with them. Indeed, H.'s death has ended the practical problem. While she was alive I could, in practice, have put her before God; that is, could have done what she wanted instead of what He wanted; if there'd been a conflict. What's left is not a problem about anything I could *do.* It's all about weights of feelings and motives and that sort of thing. It's a problem I'm setting myself. I don't believe God set it me at all.

The fruition of God. Reunion with the dead. These can't figure in my thinking except as counters. Blank cheques. My idea—if you can call it an idea—of the first is a huge, risky extrapolation from a very few and short experiences here on earth. Probably not such valuable experiences as I think. Perhaps even of less value than others that I take no account of. My idea of the second is also an extrapolation. The reality of either—the cashing of either cheque—would probably blow all one's ideas about both (how much more one's ideas about their relations to each other) into smithereens.

The mystical union on the one hand. The resurrection of the body, on the other. I can't reach the ghost of an image, a formula, or even a feeling, that combines them. But the reality, we are given to understand, does. Reality the iconoclast once more. Heaven will solve our problems, but not, I think, by showing us subtle reconciliations between all our apparently contradictory notions. The notions will all be knocked from under our feet. We shall see that there never was any problem.

And, more than once, that impression which I can't describe except by saying that it's like the sound of a chuckle in the darkness. The sense that some shattering and disarming simplicity is the real answer.

It is often thought that the dead see us. And we assume, whether reasonably or not, that if they see us at all they see us more clearly than before. Does H. now see exactly how much froth or tinsel there was in what she called, and I call, my love? So be it. Look your hardest, dear. I wouldn't hide if I could. We didn't idealize each other. We tried to keep no secrets. You knew most of the rotten places in me already. If you now see anything worse, I can take it.

So can you. Rebuke, explain, mock, forgive. For this is one of the miracles of love; it gives—to both, but perhaps especially to the woman—a power of seeing through its own enchantments and yet not being disenchanted.

To see, in some measure, like God. His love and His knowledge are not distinct from one another, nor from Him. We could almost say He sees because He loves, and therefore loves although He sees.

Sometimes, Lord, one is tempted to say that if you wanted us to behave like the lilies of the field you might have given us an organization more like theirs. But that, I suppose, is just your grand experiment. Or no; not an experiment, for you have no need to find things out. Rather your grand enterprise. To make an organism which is also a spirit; to make that terrible oxymoron, a 'spiritual animal.' To take a poor primate, a beast with nerve-endings all over it, a creature with a stomach that wants to be filled, a breeding animal that wants its mate, and say, 'Now get on with it. Become a god.'

I said, several notebooks ago, that even if I got what seemed like an assurance of H.'s presence, I wouldn't believe it. Easier said than done. Even now, though, I won't treat anything of that sort as evidence. It's the *quality* of last night's experience—not what it proves but what it was—that makes it worth putting down. It was quite incredibly unemotional. Just the impression of her *mind* momentarily facing my own. Mind, not 'soul' as we tend to think of soul. Certainly the reverse of what is called 'soulful.' Not at all like a rapturous reunion of lovers. Much more like getting a telephone call or a wire from her about some practical arrangement. Not that there was any 'message'—just intelligence and attention. No sense of joy or sorrow. No love even, in our ordinary sense. No un-love. I had never in any mood imagined the dead as being so—well, so business-like. Yet there was an extreme and cheerful intimacy. An intimacy that had not passed through the senses or the emotions at all.

If this was a throw-up from my unconscious, then my unconscious must be a far more interesting region than the depth psychologists have led me to expect. For one thing, it is apparently much less primitive than my consciousness.

Wherever it came from, it has made a sort of spring cleaning in my mind. The dead could be like that; sheer intellects. A Greek philosopher wouldn't have been surprised at an experience like mine. He would have expected that if anything of us remained after death it would be just that. Up to now this always seemed to me a most arid and chilling idea. The absence of emotion repelled me. But in this contact (whether real or apparent) it didn't do anything of the sort. One didn't need emotion. The intimacy was complete—sharply bracing and restorative too—without it. Can that intimacy be love itself—always in this life attended with emotion, not because it is itself an emotion, or needs an attendant emotion, but because our animal souls, our nervous systems, our imaginations, have to respond to it in that way? If so, how many preconceptions I must scrap! A society, a communion, of pure intelligences would not be cold, drab, and comfortless. On the other hand it wouldn't be very like what people usually mean when they use such words as *spiritual,* or *mystical,* or *holy.* It would, if I have had a glimpse, be—well, I'm almost scared at the adjectives I'd have to use. Brisk? cheerful? keen? alert? intense? wide-awake? Above all, solid. Utterly reliable. Firm. There is no nonsense about the dead.

When I say 'intellect' I include will. Attention is an act of will. Intelligence in action is will *par excellence.* What seemed to meet me was full of resolution.

Once very near the end I said, 'If you can—if it is allowed—come to me when I too am on my death bed.' 'Allowed!' she said. 'Heaven would have a job to hold me; and as for Hell, I'd break it into bits.' She knew she was speaking a kind of mythological language, with even an element of comedy in it. There was a twinkle as well as a tear in her eye. But there was no myth and no joke about the will, deeper than any feeling, that flashed through her.

But I mustn't, because I have come to misunderstand a little less completely what a pure intelligence might be, lean over too far. There is also, whatever it means, the resurrection of the body. We cannot understand. The best is perhaps what we understand least.

Didn't people dispute once whether the final vision of God was more an act of intelligence or of love? That is probably another of the nonsense questions.

How wicked it would be, if we could, to call the dead back! She said not to me but to the chaplain, 'I am at peace with God.' She smiled, but not at me. *Poi si tornò all' eterna fontana.*